Kaddour ZIANI
Waafa Arabi
Amel Sidi Iklef

Food Hygiene, Quality & Safety

Kaddour ZIANI
Waafa Arabi
Amel Sidi Iklef

Food Hygiene, Quality & Safety

ScienciaScripts

Imprint
Any brand names and product names mentioned in this book are subject to trademark, brand or patent protection and are trademarks or registered trademarks of their respective holders. The use of brand names, product names, common names, trade names, product descriptions etc. even without a particular marking in this work is in no way to be construed to mean that such names may be regarded as unrestricted in respect of trademark and brand protection legislation and could thus be used by anyone.

Cover image: www.ingimage.com

This book is a translation from the original published under ISBN 978-620-3-45118-4.

Publisher:
Sciencia Scripts
is a trademark of
Dodo Books Indian Ocean Ltd. and OmniScriptum S.R.L publishing group

120 High Road, East Finchley, London, N2 9ED, United Kingdom
Str. Armeneasca 28/1, office 1, Chisinau MD-2012, Republic of Moldova, Europe
Printed at: see last page
ISBN: 978-620-5-74305-8

Copyright © Kaddour ZIANI, Waafa Arabi, Amel Sidi Iklef
Copyright © 2023 Dodo Books Indian Ocean Ltd. and OmniScriptum S.R.L publishing group

TABLE OF CONTENTS

FOREWORD

Today we are going to talk about safety and health; every day, all over the world, people get sick from the food they eat. These illnesses are called foodborne illnesses and are caused by dangerous microorganisms and/or toxic chemicals.

There is no such thing as a "perfect" food that includes in its composition everything we need, but each food has a place and a purpose in our diet. Therefore, food is any non-toxic substance capable of ensuring the coverage of energy needs (macronutrients) and qualitative needs (micronutrients).

Although governments around the world are doing their best to improve food safety, the prevalence of foodborne illness remains a significant public health problem in all countries. The World Health Organization (WHO) estimates that 1,800,000 people die each year from diarrheal diseases, and that most of these cases can be attributed to contaminated food or water. The toll paid in human suffering is therefore enormous, especially for the most vulnerable groups (infants and young children, pregnant women, the elderly, the sick, etc.) associated with malnutrition, diarrhoeal diseases caused by unsanitary food cause unbearable devastation and are the leading cause of death among children in countries where hygiene is deficient.

The WHO/FAO International Conference on Nutrition (Rome, 1992) already stated that "...access to safe food is a universal right". Seen from this perspective, food safety must be given a high priority by governments, industry and consumers.

The WHO also recognizes that foodborne diseases are a problem for both developing and developed countries. Epidemics due to bacteria such as Campylobacter jejuni, Escherichia coli O157, Listeria monocytogenes, Samonella, etc., or sometimes viruses, have claimed thousands of lives worldwide. Every year, new risks associated with the presence of chemical contaminants or toxic products that are formed during the processing or preparation of food are discovered. Food allergies are also on the rise. This increase in cases (called "prevalence") is the result of many interacting factors, including: an increasing number of participants in the food chain, between the primary producer and the consumer; insufficiently controlled hygiene at the various stages of production and distribution, as well as at the consumer level; a change in preparation and consumption patterns: less prolonged cooking, more products consumed raw for taste or convenience, less canned but more frozen products, more product fermentations, cold smoking of fish, etc.This syllabus offers basic support to the actors of the food sector, whose main objective is to allow the involvement of the people in the field of food and to know the scientific foundations of hygiene and food safety, according to the different types of risks.The teaching of our discipline is divided into three essential sections. It aims at acquiring skills on :

1. The different food groups and origins;

2. Risks related to hazards that may be found in foods of animal origin
(biological, physical, chemical);

3. The methods of food hygiene for the control of hazards in the food industry;

4. The scientific and practical basis for official food control processes;

5. Standards (ISO 9000, 22 000) and certifications.

1. FOOD CLASSIFICATION AND TYPOLOGY

1.1 Introduction

A food is a product consumed regularly by a community, which has been able to see its safety and its benefits, it is in the long term for health or without impact on health represents a cultural heritage of inestimable value illustrating the fundamental relationship between man, nature and food. The latter's primary purpose is to ensure the coverage of energy needs (macronutrients) and qualitative needs (micronutrients).

1.2 General composition of foods

Once food is produced, selected and consumed, it is used by humans. In fact, they cannot be used as such without being transformed into nutrients in the digestive tract. In other words, between the food in its great diversity and the metabolic systems, which convert the majority of nutrients into energy to run the human body. The food designates "all the matters that it is the nature that serve or can serve for the nutrition. Food is divided into beverages and foods proper, which are essentially composed of principles of vegetable or animal origin". Nevertheless, the term "**food**" is too general, in fact many foods can belong to several food groups, so the classification of the latter can be approached according to basic criteria in many ways. All nutrients can be classified into two broad categories: macronutrients, which include carbohydrates, fats and proteins, and micronutrients, which generally include vitamins and minerals. The following tables summarize the main biological functions of nutrients in our body.

Table 1: Macronutrients

Macronutrients	Subcategory	Functions	Food containers
Glucides (= starch + glycogen + cellulose)	Slow sugars (complex)	To provide an accessible source of energy to produce metabolic reactions: ATP synthesis, glycogen storage, muscle and brain function, etc.	Pasta, rice, potatoes, dried vegetables, bread
	Quick sugars (simple)		Sugar, sweets, chocolate, white bread, fruit juice, mashed potatoes, puffed cereals, ice cream, sodas, etc.
Lipides (or fats), of animal or vegetable origin	Saturated fatty acids	(to be avoided: because it increases cholesterol and causes heart disease)	Fatty meats, whole milk, butter, cream, margarine, egg yolk, cakes, etc.
	Unsaturated fatty acids (Omega-3, Omega-6	Build up an energy reserve for long term exercise +Provide essential fatty acids (growth and functioning of membranes, (re)constitution of tissues)+ Prevent heart disease and improve performance	Olive oil, rapeseed oil, soybean oil, avocado oil, peanut oil, almond oil, cashew oil, sunflower oil and sesame oil Fatty fish (salmon, trout, sardines, etc.), nuts (+oil), sweet potatoes Corn oil, peanut oil sunflower and sesame seeds
Protein	Animals	(Minor energy role)	Meat, fish, eggs, dairy products
	Vegetables	Cell formation, muscle development and function + Physiological role (regulation of hormones, biochemical reactions) + Immune protection + Hemoglobin transport	Cereals, legumes, pulses, oilseeds, soybeans

Table 2: Main vitamins in food

Vitamin	Functions	Food containers
A	Synthesis of visual pigments Maintenance in good condition of the mucous membranes and the skin	Butter, milk, eggs, liver, vegetables, fruits
B1 (Thiamine)	Blood sugar balance Glycogen storage Lipid synthesis Decreased recovery time Cramps disappear	Wholemeal bread and cereals, pulses, meat (liver), eggs, nuts
B6 (Pyridoxine)	Metabolism of carbohydrates, lipids, proteins	Egg yolk, liver, yeast, soy
B12 (Cyanocobalamin)	Antianemic cell construction Protein metabolism	Meat, liver, egg, fish
C	Connective tissue health Immune defences Fatigue resistance Iron absorption	Fruits, green vegetables
D	Calcium and phosphorus absorption	Cod liver, eggs, butter, times, fatty fish
E	Cellular development and function (energy creation)	Cereals, oils (rapeseed, corn, soybean, olive, sunflower)
K	Allows coagulation	Liver, eggs, vegetables green
PP	Cellular metabolism Hydrogen transfer Fatty acid synthesis	Meat, fish, dairy products, eggs, legumes

Table 3: Minerals

Mineral	Functions	Food containers
Sodium	Transmission of nerve and muscle impulses Maintenance of cellular balance	Salt
Potassium	Fluid control Muscle and nerve functions	Fruits, vegetables, cereals
Phosphorus	Bioenergetics	Meat, fish, eggs, milk, cereals, vegetables
Magnesium	Maintenance of neuromuscular balance Protein synthesis	Vegetables (lentils), meats, dried fruits, cereals
Calcium	Muscle contraction Nerve conduction Rhythmic activity of the heart Bone formation Blood coagulation	Milk, cheese, green vegetables, fresh and dried fruits
Iron	Formation of red blood cells Oxygen transport	Meat, eggs, milk, pulses, pasta, wine, whole grains
Zinc	Degradation of carbohydrates, lipids, proteins Cell growth Immunity	Eggs, fish, meat, products dairy products, lentils, nuts, shellfish
Sulfur	Production of ATP Elimination of toxins	Green and dry vegetables, meats, eggs, cheese
Copper	Redox reactions	Nuts, shellfish, dried vegetables, potatoes
Fluorine	Bone strength	Toothpaste, chewing gum
Iodine	General metabolic activity Neuromuscular and circulatory functions	Shellfish, fish, salt, fruits and dried vegetables
Cobalt	Oxygen transport	Girolles, food rich in B12
Chrome	Lipid metabolism	Shellfish, fruit, beer

The constituents of a foodstuff are very numerous and very diversified as well in their chemical structures as in their proportions:

1. **Main:** water, lipids, carbohydrates, proteins;

2. **Minors:** mineral elements, vitamins, dietary fibers, flavors (natural), colorants (natural), secondary metabolites of plants (phenols).

3. **Additives:** colorants, preservatives, etc.

4. **Natural toxins:** cyanogenetic glucosides, solanine (cholinesterase inhibitor), anti-vitamins (Avidin of egg white anti-vit H/vit B8);

5. **Degradation products:** biogenic amines (histamine), Maillard reaction products, etc.

6. **Foreign** substances phytosanitary agents (residues of antibiotics, mycotoxins, bacterial toxins, etc.

7. **Organisms:** parasites, molds, bacteria, viruses, prions, etc.

1.3 Different food groups

Although there is no essential food, there are however, due to diversity, differences and similarities between foods. It is the notion of class or groups of foods that has led to the need to group them together and therefore to a classification. However, even if any classification has a chemical (nutritional) justification, it also has an educational purpose. This is why there are several classifications, all of which are imperfect, but each of which may have its own interest (and educational limits) depending on the objectives.Thus, nutrition education in underprivileged areas or in developing countries development is based on a classification into three groups proposed by the WHO by For example, building foods (those that provide protein, such as meat and pulses); energy foods (including starchy foods and fats); and protective foods (rich in micronutrients such as fruits and vegetables). We had developed, with educational success, a classification into nine groups: meat, deli meats, eggs and fermented cheeses to emphasize a commonality in terms of saturated fatty acids; fish and fish products to emphasize their specific content of long-chain omega-3 fatty acids; fruits and vegetables including potatoes to emphasize a similarity in terms of phytomicroconstituents, micronutrients and fiber; cereals, bread and legumes to emphasize their richness in vegetable proteins; milk and dairy products (except fermented cheeses); vegetable fats and butter; oleaginous fruits, which should be kept separate; sweet foods and beverages to emphasize their high content of simple sugars; and alcoholic beverages and alcoholic beverages to emphasize the specificity of alcohol.The "official" classifications (PNNS) are simple, even simplifying, but accepted with seven groups: meat and related products, eggs, fish; dairy products; fruits and vegetables; starchy foods, potatoes, bread, cereals, pulses; fats; sweet products and beverages.

2. FOOD HYGIENE MAINTENANCE

2.1 Introduction

Food security and food hygiene are not to be confused with food hygiene and food safety! These terms are misused in common parlance: although **food** security is an expression that refers to the security of food supplies in quantity and quality (food self-sufficiency). Food security has four pillars:

a. Access: the ability to produce one's own food and therefore have the means to do so, or the ability to purchase one's own food and therefore have sufficient purchasing power to do so,

b. **Availability:** sufficient quantities of food, whether from domestic production, stocks, imports or aid,

c. **Quality:** of food and diet from nutritional and sanitary points of view,

d. **Stability:** of access capacities and therefore of prices and purchasing power, of availability and quality of food and diets.

While, **food** safety is the assurance that food does not harm the consumer when it is prepared and / or consumed in accordance with its intended use. Examples of means implemented for food safety: control of origins, control of composition, detection of sources of bacterial contamination, control of the manufacturing or processing chain and control of the cold chain.

However, **food hygiene** (dietetics) is a medical term referring to the reasoned choice of food (the rules of nutrition and dietetics) and **food hygiene** (Food Hygiene) refers to all the conditions and measures necessary to ensure the safety and wholesomeness of food at all the stages of the food chain, (definitions from the NF V 01-002 standard on food hygiene).The risk of food poisoning for the consumer remains low but would still be largely reduced by the application of simple and effective hygiene rules. These rules of hygiene concern of course the producers, the distributors but also each individual.

Food hygiene ensures the safety and wholesomeness of food.

In other words, **food safety** "ensures that food, when consumed in accordance with its intended use, is acceptable for human consumption. The concept of safety is different from that of security. It applies more to the intrinsic characteristics of the product, i.e. taste, odour, texture, etc.

Ces deux composantes de l'hygiène sont indissociables (*cf.* Figure 1)

The hygiene of food is composed of several areas all equally important: the hygiene of the staff, the hygiene of the premises (cleaning, disinfection, etc.), the conditions of storage, handling, transport, etc. All these points where hygiene is crucial are included in the so-called "5 M method" or "**Ishikawa diagram**" (see HACCP "Hazard Analysis Critical Control Point").

2.2 Agro-Industrial Hygiene

In the **food** industry, the hygiene measures that allow us to obtain healthy food **have** two components: **Safety:** food without danger (no salmonella, no bits of glass, etc.) and **Healthiness:** acceptable food, consumable (no bad smell, no alteration, etc.).

2.2.1 The importance of agro-industrial hygiene

▪ Obvious sanitary importance: "Less hygiene = More sick people";

▪ Economic importance: preservation of the product, possible exports, less

A hygiene error is often the death of the company";

▪ Legal importance: the Algerian official journal imposes hygiene rules.

2.3 Hygiene control in the food industry

Food contamination can occur at any stage of food preparation and storage. Unfortunately, the majority of food poisoning is due to negligence or lack of knowledge of the rules and precautions necessary for the preparation and preservation of food.

2.3.1 Good Hygiene Practice Guide (GHPG)

C'est Simple

The GBPH is a reference document, of voluntary application, evolutionary, conceived by the professionals for the food sector. Each GBPH gathers the specific recommendations of the sector to which it refers. These documents recommend means, adapted methods, procedures whose implementation must lead to the control of the sanitary requirements (regulatory or not) during the preparation, the transformation, the manufacturing, the packaging, the storage, the transport, the distribution, the handling and the sale or the availability of the foodstuffs to the consumer. They also allow professionals to harmonize the hygiene rules for a sector of activity.

The GBPH recommendations have been validated from a **Scientifique** and **Réglementaire** perspective in order to ensure the safety and wholesomeness of the products. The professionals can be led to implement only a part of the proposed control measures or to choose other means which allow to reach the technical and regulatory objectives of safety and wholesomeness of the products; in this case, they will have to demonstrate that the implemented means are relevant and effective. This guide also describes the application of the principles of hygiene in its activity, starting from the analysis of the potential food risks of an operation and the collection of the various means of control and monitoring at each point of risk (the GBPH is a kind of HACCP plan for a family of products, for companies of the same sector). As the GBPH is specific to a sector, it gives precise details for the products of the sector: instead of general and abstract texts, the GBPH contains clear and detailed instructions, it constitutes a reference of the good hygienic practices for the integration of the new professionals who start in the activity.The GBPH also includes a common part for all sectors of activity, where the common hygiene provisions concerning the premises, the equipment, the personnel, the water, the air, the waste, etc. are reminded. Each professional chooses only one or another of the means proposed by the guide, according to the specific conditions of the **"5M" of** his operation. He then constitutes his own "doctrine" in terms of hygiene by writing a company reference system, which is based on the GBPH and adapted to his particular case.

2.3.2 The Great Principles of Hygiene

Hygiene measures include: room hygiene, equipment hygiene, personal hygiene and food hygiene. This course covers the principles of hygiene, and develops the major causes of "non-hygiene" using the **"5M"** diagram.

Hygiene is above all a question of rules to be observed and an indispensable education.
a. **Material**
A poor-quality foodstuff on arrival will be a drag that the LPN will drag all the way to the consumer's table; the reception of raw materials is therefore a key position in the LPN. With regard to the material, the origin, safety, labelling and temperature of the products are checked. The hygienic conditions that govern the production of foodstuffs are fundamental to ensure good hygiene in the rest of the food chain. It is therefore important to avoid production in areas where the

environment may pose a threat to food safety. This means controlling air, water and soil contamination.It is important to ensure good quality feed for livestock, poultry, fish or any other livestock. The health of the animals must be ensured, as well as the veterinary practices used. Fertilizers, plant protection products and pesticides should be used with care to ensure that the plants are healthy and safe for humans or animals that eat them.At the time of slaughter or harvest, all plants, grains, fruits, etc., or animals obviously unfit for consumption must be eliminated. The waste resulting from these operations must be disposed of hygienically. Crops and meat products must be protected from all possible contamination (pests, microbial germs, undesirable chemical or physical agents) that may occur during the handling, storage and transportation of foodstuffs. The people involved in these primary production operations (producers, fishermen, farmers, staff in slaughterhouses, in silos, etc.) must demonstrate the best possible personal and body hygiene. Of course, they must be informed of the issues at stake and of the risks they would run. Their responsibility must be engaged as soon as they are trained and/or informed.

Rules

> Dans tous les cas, et quel que soit le mode d'approvisionnement en matières premières et ingrédients, il important de choisir judicieusement les matières premières et les ingrédients auprès de commerçants connus du marché et offrant des garanties de salubrité.

b. Material

Equipment includes machines, tools, tables, conveyors, bins, etc. All work surfaces must be kept clean, the equipment must be designed to prevent the accumulation of dirt and allow for easy and thorough cleaning.

Ces règles s'appliquent à tout ce qui entre en contact avec les aliments

c. Environment

Reducing the risk of hazardous food manufacturing is done by taking preventive measures to ensure the safety and wholesomeness of food. Control of hazards is achieved through control procedures at all critical stages of food manufacturing and processing operations. However, it is not enough to have control procedures in place; it is also necessary to ensure that these procedures remain effective and appropriate over time.

C.1 Hygiène de l'environnement

The factory or workshop must therefore be far from sources of contamination. For example: minimum distance from a road = 5 m, a house = 50 m, a livestock= 100 m, a waste stockpile = 200 m). NB: The factory itself can be a source of nuisance for the neighborhood = specific surveillance is required.

C.2 Hygiène des bâtiments et locaux

C.2.1Rationaldevelopments

This layout is based on easy-to-clean shapes (smooth walls, non-slip floors, no sharp corners, but rounded grooves (wall-ground), no corners, no slope of the floor > 1%, no dust nests: cables, shelves, etc.), as well as a wide space between wall and equipment, and around each machine on sealed feet. The air must be controlled to ensure the prevention of contamination, there are two complementary aspects: Renew the indoor air to eliminate endogenous contamination (fumes, smoke, aerosols, food particles, etc.) and filter the outdoor air to eliminate dust and bacteria. This requires a ventilation/filtration unit.

Separate sectors

Incompatible" areas must be physically separated. It should not be possible to move directly from a contaminated area to a clean area (e.g., a clean room). (e.g. receiving/manufacturing). Same thing between hot and cold areas, i.e. group cold areas together and the same with hot areas and identify areas with "mandatory" temperatures.

Identify the circuits: Forward motion is imperative.

The circuit must not include any backtracking or crossing. We go from dirty to clean, to avoid cross-contamination (e.g. raw materials do not cross the processed product). The separate and forward sectors are normally shown on the plant layout.

d. Methods

An automated operation is less risky than a manipulation. But also mechanical operations make the whole product accessible to a contaminant (example: slicing, chopping, grinding, mixing, etc.). We will be very attentive to the cleanliness of the machines (cleaning and disinfection). The hardest jobs are the places hygiene faults, we try to reduce drudgery, with the participation of the worker (ergonomics).Ergonomics: the word "ergonomics" comes from the Greek ergon (work) and nomos (laws, rules). Ergonomics can therefore be defined as: "the science of work with the aim of adapting work to man, in a way that allows the implementation of scientific knowledge relating to man and necessary to design tools, machines and devices that can be used by the greatest number of people with the maximum comfort, safety and efficiency.

e. Hands of work

Hands are a particularly important vector for the transmission of pathogenic microorganisms causing food poisoning. The control of the hygiene of the "Hands of work" is very important, indeed, they condition the others: they control the raw materials ( ), they clean the equipment ( ), they implement the environment ( ) (eg. separate sectors), they make the methods ( ); and they ( ) are considered a major source of germs (1011 bacteria/g). Therefore, LPN personnel must be clean, healthy, trained in hygiene, and trained in the proper use of their position.

• Personal Hygiene (Cleanliness)

The hygiene of the personnel who are in contact or can be in contact with food must be perfect, in order to avoid contamination and transmission of diseases to consumers. Personal hygiene obviously concerns the cleanliness of the body, hands and hair, but also the cleanliness of work clothes (clothing, headgear, shoes, masks, gloves, etc.). A certain number of behaviors or activities are contrary to the rules of food hygiene such as smoking, spitting, coughing, sneezing, drinking or eating near food being processed.

L'utilisation de gants ne dispense pas du lavage des mains

- Personal Hygiene (Health condition)

Any person suffering from or carrying a disease that may be transmitted through food, or suffering from, for example, infected wounds, skin infections or lesions, or diarrhea, shall not be permitted to handle food or enter any food handling area in any capacity where there is a risk of direct or indirect contamination of food.

En pratique, agir comme si chacun était porteur sain

2.4 Cleaning & Disinfection in the food industry

The notions of Cleaning and Disinfection are defined in the **NF V 01- 002, 2008** standard as follows:Cleaning is the removal of soil, food residues, dirt, grease or other objectionable matter; while Disinfection is the action of reducing by chemical agents or physical methods the number of microorganisms to a level that will not compromise the safety or suitability of food. Cleaning and disinfection programs must be established to ensure that food manufacturing equipment and environment are maintained in a satisfactory state of hygiene. These programs must be monitored to ensure their continued adequacy and effectiveness.

Une désinfection ne doit être réalisée qu'après un nettoyage correct.

Cleaning is an operation of maintenance and upkeep of the premises and equipment whose main objectives are to ensure a pleasant appearance (notion of comfort) and a level of cleanliness (concept of hygiene) and to keep the food environment healthy from the moment the goods are received to the consumer.

This operation of elimination of dirt: biological, organic or mineral is carried out by a process respecting the state of the treated surfaces and calling upon, in variable proportions, the following combined factors: Chemical action, Mechanical action, Time of action and Temperature.

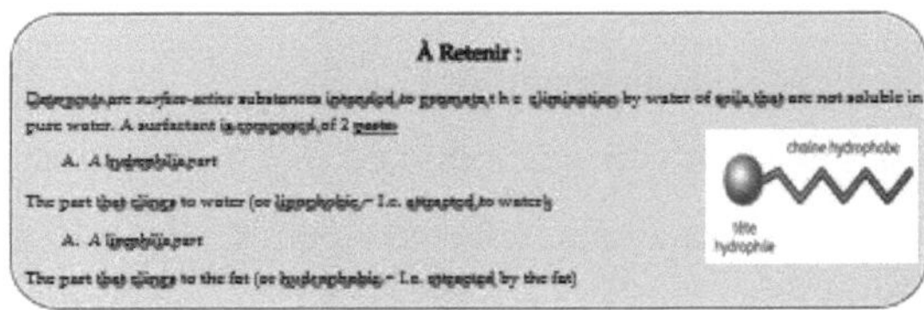

The presence of these four factors (is essential and their combination is variable Whatever the method implemented and the organization chosen, they are always present and the decrease of one is always compensated by the increase of one or more of the others:It is the action of a detergent product, the choice of products must be adapted to:

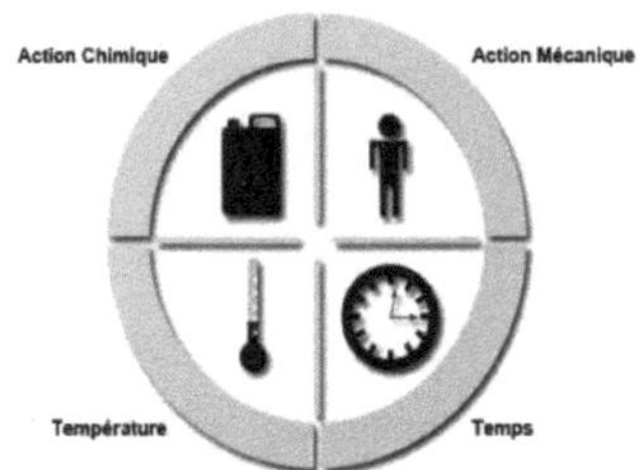

1. Concentration (too low: bad result; too high: more difficult rinsing, risk of traces) ;

2. Detergent characteristics (foaming, soaking, etc.);

3. The nature of the soil (Organic = alkaline detergents; Mineral = acid detergents);

4. The constitution of supports à clean (ease of decreasing: glass>steel>aluminum>rubber >plastic > ...> wood) ;

5. Compatibility with production (absence of residues);

6. Operator safety (products with minimum risk).

It is the action of rubbing or using equipment such as single brushes or scrubbers, it has for main purpose to resuspend the dirt and eliminate them. This improves the effectiveness of detergents. It is a very important factor because it intervenes at all stages of the cleaning process to put the surface to be cleaned in constant contact with fresh solution and to create the forces necessary for the removal of the soils.

Its role in detergency is far from negligible, the temperature :
• It lowers the surface tension, so that it has been said that hot water is already a detergent;
• Accelerates most chemical reactions and in particular, saponification and

hydrolysis;
• Softens oils, greases and waxes and thus facilitates the penetration of the detergent;

Finally, it was found that when the temperature was increased by **12°C,** the cleaning or disinfection speed **increased** by **2 times**.

However, the temperature increase has limits:

I. High temperature water boiling ;

II. High cost of thermal energy;

III. Low resistance thermal resistance of some materials (plastics plastics, rubber, etc.);

IV. Coagulation temperature of certain stains (albuminoids);

V. Selected application method: For example, the dermal resistance in manual application does not exceed 50°C.

VI. Iodinated acids, for example, cannot be used above 43°C, otherwise sublimation of the iodine will occur; chlorinated alkalis should be used at temperatures below 72°C.

The approximation of the cleaning kinetics to a first order reaction and is not based on any serious assumptions.Caution: prolonged application time of some chemicals can cause deterioration of the substrate.

Toujours penser que le nettoyage est un mix de ces facteurs.

Disinfection is an operation with a momentary result allowing the elimination or killing of microorganisms and/or the inactivation of undesirable viruses, carried by contaminated inert media (Surfaces; Objects; Medium (water, air)), according to the fixed objectives. The result of this operation is limited to the microorganisms present at the time of the operation" (Agence Française de Normalisation, **AFNOR, 1981**).Disinfection is used to prevent cross-infection and to achieve the lowest levels of contamination in the environment. Disinfection has for to reduce the microorganisms present on objects, surfaces, etc.

La désinfection = élimination de toute souillure MICROBIENNE.

The main factors influencing the effectiveness of disinfectants are:

a. Contact time

For each disinfection process, a contact time is required between the product and the microorganisms: Short: vegetative bacteria, Long: Koch's bacillus, hepatitis virus, etc.

b. Temperature

Disinfection is always faster when the temperature is higher. In some cases, the product is diluted in hot water. Ex: Alcohol: it takes 10 minutes at 30°C to kill a population, 300 minutes at 20°C to kill the same population!

c. Product concentration

A disinfectant that is too concentrated will cause surface coagulation of organic matter and thus prevent the product from penetrating deeply, becoming irritating, corrosive and unnecessarily expensive. The recommended dilutions must be strictly respected. Conversely, a product is less active when it is too diluted. Excessive dilution sometimes leads to a dramatic reduction in activity.

d. pH

Some products are more active in an acidic environment, others in an alkaline environment.

e. Inhibitors

The action of disinfectants can be inhibited by a variety of substances, when the dilution water is hard, calcium moderately inhibits most disinfectants.

3. HACCP

3.1 Hazard control: HACCP

HACCP is the well known acronym of "Hazard Analysis Critical Control Point". In French, it is a system of hazard analysis and critical control points. This method has become synonymous with food safety worldwide. The HACCP concept was originally developed as a microbiological safety system at the beginning of the U.S. space program in the 1960s to ensure the safety of food for astronauts. The original system was designed by Pillsbury Company, in cooperation with the National Aeronautics and Space Administration (NASA) in the United States and the US Army Laboratories.

Here are some basic terms that are essential to understanding:

ISO 22 000 : 2005

§ HACCP (analyse des dangers ; points critiques pour leur maîtrise)

Démarche qui identifie, évalue et maîtrise les dangers significatifs au regard de la sécurité des aliments.

The HACCP method makes it possible to analyze and control the dangers:

ISO 22 000 : 2005

§ Dangers

Agent biologique, chimique ou physique, présent dans un aliment ou état de cet aliment pouvant entraîner un effet néfaste sur la santé.

The key to this method is the identification and control of critical points. It is important to define this notion of critical point which is often misunderstood.A critical point is a crucial point that must be mastered to avoid negative effects later on.

ISO 22 000 : 2005

§ Point critique pour la maîtrise (CCP)

Étape à laquelle une mesure de maîtrise peut être exercée (et est essentielle) pour prévenir ou éliminer un **danger** menaçant la **sécurité des aliments** ou le ramener à un niveau acceptable.

HACCP is based on the principle that food safety hazards can be either eliminated or minimized through prevention at the production stage rather than through inspection of finished producs. Its objective is to prevent hazards as early as possible in the food chain. The HACCP method can be applied from primary production to consumption. Companies using HACCP are able to provide better assurances about food safety to consumers and food regulatory authorities.

3.2 Principles of HACCP

The HACCP includes seven principles, which allow to establish, implement and conduct a HACCP plan:

▪ **Principle 1**

- Conduct a hazard analysis;

- Identify potential hazards associated with all stages of production;

- Assess the likelihood of occurrence and the severity of the effects of each hazard.

▪ **Principle 2**

- Identify critical control points (CCP);

- Determine the stages at which monitoring can be performed and is essential to prevent or eliminate a food safety hazard.

■**Principle 3**

- Set the critical threshold(s). The critical limit is the criterion that distinguishes acceptability from non-acceptability. They must involve a measurable parameter and can be considered the absolute threshold or safety limit for CCPs.

■**Principle 4**

- Implement a monitoring system to control CCPs at the by means of tests or planned observations.

■**Principle 5**

- Determine corrective actions to be taken when monitoring indicates that a particular CCP is not under control.

■**Principle 6**

- Apply verification procedures to confirm that the HACCP system is working effectively.

■**Principle 7**

- Create a file that includes all procedures and records regarding these principles and their implementation.

3.3 HACCP steps

The application of HACCP principles consists of the following tasks, as described in the logical sequence of application of HACCP.

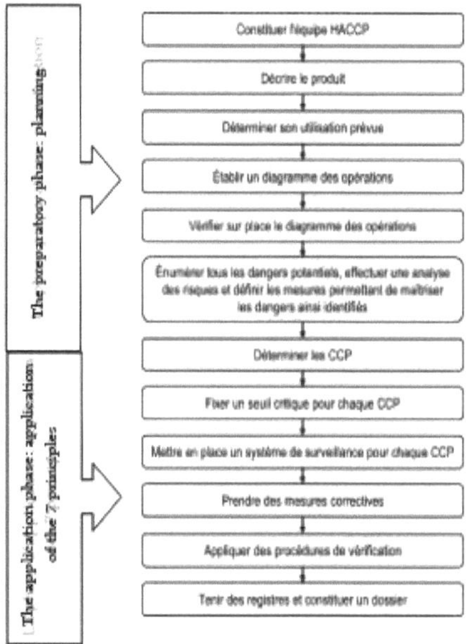

Figure 2: Logical sequence of HACCP application

3.3.1 Step 1: Assemble the team HACCP

It is really important that the implementation of HACCP is not the work of an isolated quality manager but that it is the work of a multidisciplinary team: the food safety team. Ideally, this team should be composed of people from different functions of the organization. It is possible, when the need arises, to call upon external experts (microbiologist, consultant, supplier for example).

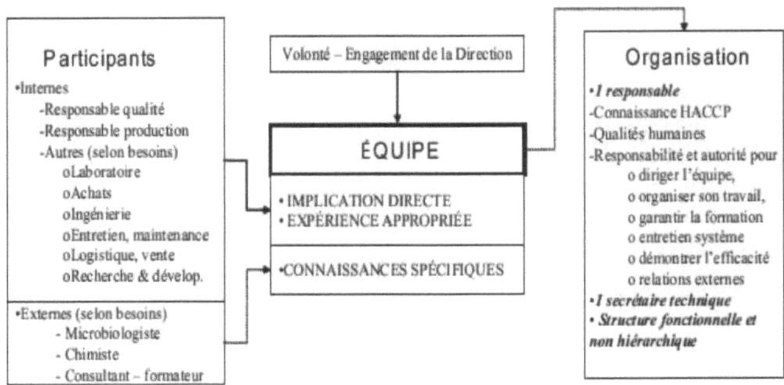

3.3.2 Step 2: Describe the product and its distribution

This description is primarily about the finished product.The characteristics of the finished products shall be documented to the extent necessary to perform the hazard analysis, containing information relating to the following, as appropriate:

a. Product name or similar identification;

b. Composition;

c. Biological, chemical and physical characteristics relevant to food safety;

d. Expected shelf life and storage conditions;

e. Packaging;

f. Food safety labeling and/or instructions for handling, preparation and use;

g. Distribution methods.

3.3.3 Step 3: Identify the intended use of the product

This step completes the previous one: it leads to the formalization of the conditions of storage, distribution and use of the product by the end user, who is either the consumer or the processor using the product as an ingredient. It is necessary to foresee all the "normal" uses of the product: (Temperature of conservation; Thermal treatment (cooking or reheating); The shelf life of the product (DLC or DLUO); The mode of use of the product).

3.3.4 Step 4: Build the process diagram

The main purpose of a flow diagram is to enable the identification of the possible occurrence, introduction or increase in levels of hazards that cannot be identified in the other initial steps. Where appropriate and necessary for hazard identification, hazard assessment and control measure evaluation, additional flow diagrams/schematics or facility descriptions may be developed for flows other than products (such as air flow, personnel flows, equipment flows, supplies, etc.) to indicate the relative location of other control measures and to indicate the possible occurrence or transfer of food safety hazards.

3.3.5 Step 5: Confirm the diagram on site

Based on the documents produced (process and flow diagrams), the HACCP team must go and confirm all this information in the field. For the realization of this verification, it is advisable to follow the forward movement of the product: from the reception of the raw materials and the ingredients until the expedition of the finished product. Finally, it is an opportunity to correct any errors made during the construction of the diagram or any deviations from the information collected.

3.3.6 Step 6: Hazard Analysis & Control of identified hazards

The hazard analysis must be performed for all products (or category), existing or new. Indeed, changes in raw materials, formulations, treatment and preparation processes, packaging, distribution and/or use of the product will require a revision of the hazard analysis. The hazard analysis includes the following major actions:

- Identify hazards;

- Assessing hazards;

- Define and implement control measures (corrective actions).

a. Identification of hazards

All reasonably foreseeable food safety hazards related to the type of product, type of process and processing facilities used shall be identified and recorded. The acceptable level of hazard for the finished product shall, to the extent possible, be determined for each identified food safety hazard.

b. Assessment of hazards

A hazard assessment must be carried out to determine, for each identified food safety hazard, whether its elimination or reduction to acceptable levels is essential for the manufacture of safe food and whether its control is necessary to achieve the defined acceptable levels.The term hazard is therefore not to be confused with the term risk which, in the context of food safety, refers to a function of the probability of an adverse health effect (e.g., contracting a disease) and the severity of that effect (death, hospitalization, missed work, etc.) when the subject is exposed to a specific hazard.

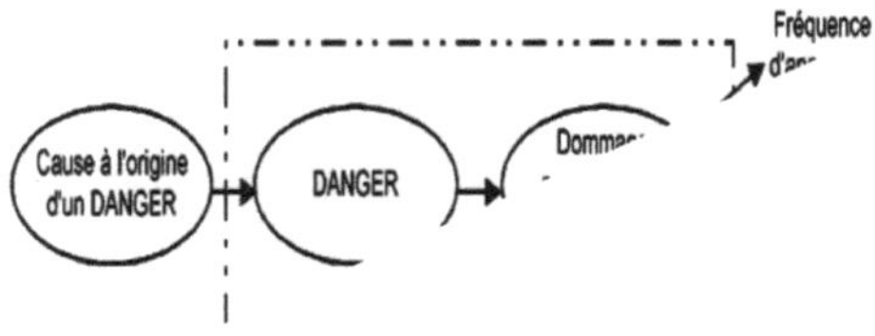

Figure 3: Hazard assessment scenario

c. Determination of control measures

The food safety team must determine the control measures to avoid, reduce to an acceptable level or eliminate previously identified hazards, especially at sensitive stages.

ISO 22 000 : 2005

§ Mesure de maîtrise

Action ou activité à laquelle il est possible d'avoir recours pour prévenir ou éliminer un danger lié à la sécurité des denrées alimentaires ou pour le ramener à un niveau acceptable.

§ Mesure corrective

Toute mesure à prendre lorsque les résultats de la surveillance exercée au niveau du

CCP indiquent une perte de maîtrise.

3.3.7 Step 7: Determine Critical Control Points (CCP)

ISO 22 000 : 2005

§ Maîtrise de CCP

Stade auquel une surveillance peut être exercée et est essentielle pour prévenir ou éliminer un danger menaçant la salubrité de l'aliment ou le ramener à un niveau acceptable.

The determination of a CCP within the HACCP system can be facilitated by the application of a **decision tree** (example of a decision tree) that provides logic-based reasoning. **Flexibility in the** application of the decision tree is required, depending on whether the operation involves production, slaughter, processing, storage, distribution, etc. It should be used as a guide when determining CCPs. If a hazard has been identified at a step where a safety control is required and no control measure exists at that or any other step, then **the product or process** at that step, or at an earlier or later stage, should be **modified to** provide a **control measure**.

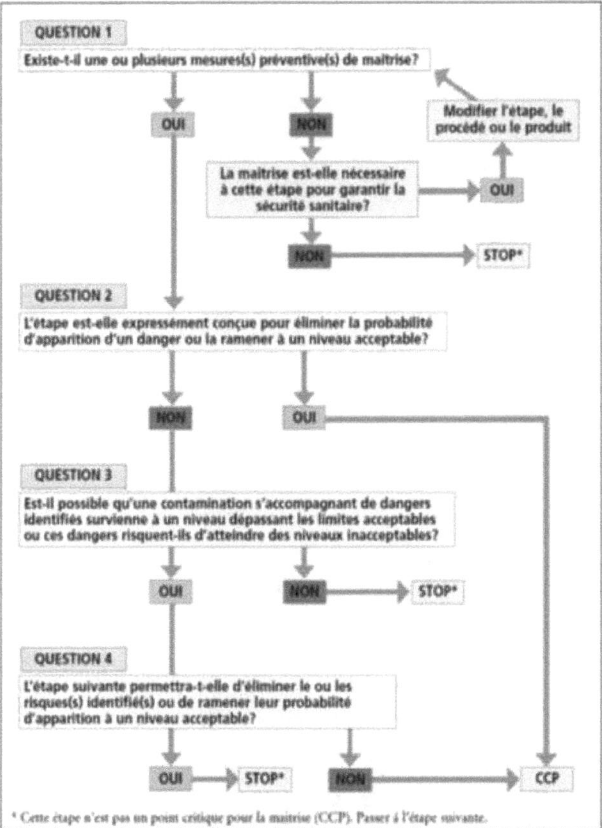

Figure 4: Decision tree for determining critical points for control

3.3.8 Step 8: Set critical thresholds for each CCP

Thresholds corresponding to each of the critical points for the **control of hazards** must be **established** and **validated** if possible. A **critical threshold** is defined in particular as a criterion for defining **acceptable** and **unacceptable** levels. A critical limit represents the limits used to judge whether an operation results in safe products following the correct application of preventive measures.

In some cases, several **critical thresholds** are set for a given stage. These include temperature, time, moisture content, pH, percentage of free water, as well as organoleptic parameters such as visual appearance and consistency.

> ISO 22000 : 2005
>
> § Limite critique
>
> Critère qui distingue l'acceptabilité de la non acceptabilité.

Note: Do not confuse the concepts of critical limits and acceptable levels of hazards. These two elements are obviously linked but are different. Indeed, to guarantee that the hazards in the product will be below the acceptable levels, it will be necessary to guarantee the control of the CCP thanks to the critical limits.

Example: Product: Milk

Acceptable level of biological hazards: absence of salmonella

CCP : Pasteurization

Control measure: respect of the time/temperature pair

Critical limit: 72°C for 2 min.

3.3.9 Step 9: Implement a monitoring system for each CCP

ISO 22000 : 2005

§ Surveillance

Action de procéder à une séquence programmée d'observations ou de mesurages
afin d'évaluer si les mesures de maîtrise fonctionnent comme prévu.

For each CCP, a monitoring system must be established to demonstrate that the CCP is under control. This system shall include all scheduled **measurements** or **observations** related to the critical limit(s). The monitoring system shall consist of relevant procedures, instructions and records covering the following:

a. Measurements or observations providing results within an appropriate time interval;

b. Monitoring devices used;

c. Applicable calibration methods;

d. Monitoring frequency;

e. The responsibility and authority for monitoring and evaluating monitoring results ;

f. Registration requirements and methods.

The monitoring methods and frequency must allow for early identification of exceedance of critical limits in order to isolate the product prior to use or consumption.All statements and reports resulting from **CCP monitoring** must be signed by the person(s) responsible for the monitoring operations, as well as by one or more company officials.

3.3.10 Step 10: Determine corrective measures (Establish corrections and corrective actions)

Corrections and corrective actions must be implemented as soon as a **critical limit** is exceeded and/or when a critical point is no longer under control.

ISO 22000 : 2005

§ Correction

Action visant à éliminer une non-conformité détectée.
  § Action corrective
Action visant à éliminer la cause d'une non-conformité détectée ou d'une autre situation indésirable.

NOTE: A correction should reduce the severity of perceived (during monitoring) undesired effects. The correction concerns the fate of potentially hazardous products.

3.3.11 Step 11: Establish audit procedures

This step is intended to determine whether HACCP is working properly and possibly to identify defects that need to be corrected. **Verification** and **auditing methods**, procedures and **tests** may be used, including random sampling and analysis.

§ Correction

Action visant à éliminer une non-conformité détectée.
  § Action corrective
Action visant à éliminer la cause d'une non-conformité détectée ou d'une autre situation indésirable.

3.3.12   Step 12: Build records and keep logs

Accurate and thorough **record keeping** is essential to the implementation of **HACCP. HACCP procedures** should be **documented** and should be appropriate to the nature and scale of the operation.

Table 4: Objectives and purpose of the different HACCP documents

Objectif	Finalité
The HACCP plan	
Determine the elements of control of the critical control points (stage, hazard, control measure, critical limit, monitoring, corrective action, register)	To guarantee the effective control of the sanitary safety of the products made by the organization
HACCP procedures	
Provide information on how to	Answer the questions: Who does what
Of realize the different activities	? How? Where? When? Define the
(reception control, CCP monitoring,	responsibilities and the areas
calibration, withdrawals, corrective actions,	application in case of choice or
etc.)	decision to be made
HACCP instructions	
Provide information on how to	Answer the questions: Who? and
to carry out the various stages of the	How ? which limit the choice.
the product's manufacturing process, the	The instruction should be simple and
procedures	directive
HACCP records	
Demonstrate the execution of an activity, procedure, instruction or other	Provide indisputable proof of the completion of the activity including an audit.

The HACCP is not a standard in the true sense of the word, it is a method or an approach that allows the establishment of a system that aims, in the case of food, the production of a safe food, and this by controlling the hazards that are unacceptable and can harm the health of the consumer. For the HACCP system to be effectively implemented, it is essential to train To help develop specific training in support of HACCP, instructions and work procedures should be formulated that clearly define the different tasks of operators at each of the points of contact. In order to contribute to the development of specific training in support of the HACCP system, instructions and work procedures should be formulated that precisely define the different tasks of the operators at each of the critical control points.

4. HYGIENE MEANS

QUALITY AND STANDARDS

4.1 The means of hygiene: the search for quality

The appreciation of a product by an individual leads to the formulation of a value judgment which, for this individual, will be the "quality" of the product, without this evaluation necessarily being shared by all. In a global way, and since the evaluation of a product can vary from one individual to another, quality can be defined as **"the ability of a product or a service to satisfy the needs of users" (AFNOR, 1982)**. In the realization of the product and in the internal relations within the organization, the various actors must therefore focus on the product expected by the customer.

4.2 Components of quality

For a food product, quality has main components: (**4S**)

a. *Sécurité (composante hygiénique)* : We want less risk

Its purpose is to guarantee the safety of the product to be consumed (microbes, toxins, chemical pollutants or foreign bodies, when they are in food, can make us sick or even be fatal);

b. *Santé (composante nutritionnelle)* : We want more assets

The food to be consumed must be dietetic; not to degrade our physical integrity but on the contrary to maintain and improve our health.

c. *Saveur (composante organoleptique)* : We want to please ourselves

Its objective is to satisfy the 5 senses (especially color, smell, taste, etc.).

d. *Service (composante d'usage)* : We want it to be convenient

S'ajoute à ces 4S les 2R :

Consumers are looking for better value for money;

e. *La régularité (composante constante)* **: We don't want any surprises**

Quality must be reproducible;

f. *Technologie (composante technologique)*   *:*

Expectations of other users (transformer(s), distributor(s), etc.).

4.3 ISO 9000" standard

ISO (International Organization for Standardization) is a worldwide federation of national standards bodies. The development of international standards is generally entrusted to ISO technical committees.This International Standard provides the fundamental concepts, principles and vocabulary of quality management systems (QMS) and serves as the basis for other quality management system standards. This International Standard is intended to help the user understand the basic concepts, principles and vocabulary of quality management so that they can effectively and efficiently implement a QMS and create value from other management system standards. ISO 9000: 2015 describes the fundamental concepts and principles of quality management that are applicable to all of the following entities:

• Organizations seeking sustainable performance through the implementation of a quality management system ;

• Customers seeking assurance of an organization's ability to consistently deliver products and services that meet their requirements;

• Organizations seeking to ensure that their supply chain will meet their product and service requirements;

• Organizations and stakeholders seeking to improve communication through mutual understanding of quality management vocabulary;

• Providers of quality management training, assessment or consulting;

• Bodies responsible for assessing conformity to the requirements of ISO 9001 ;

This international standard encourages the adoption of a **process** approach when developing, implementing and improving the effectiveness of a quality management system, in order to increase **customer satisfaction** by meeting their **requirements**. For an organization to operate effectively, it must identify and manage many interrelated activities. Any activity that uses resources and is managed in such a way as to enable the transformation of inputs into outputs can be considered a process. The output of one process is often the input to the next process. Process: refers to the set of correlated or interacting activities that use inputs to produce an expected result. When used in a quality management system, this approach emphasizes the importance of :

a) Understand and meet the requirements;

b) Consider processes in terms of added value;

c) Measuring the performance and efficiency of processes;

d) Continuously improve processes on the basis of objective measurements.
In addition, the concept of the "Demming Wheel" applies to all processes. The Demming Wheel can be succinctly described as follows:
Plan: Establish the objectives and processes necessary to deliver results that meet customer requirements and organizational policies.

Do: implement the processes.
Verify: monitor and measure processes and product against policies, objectives and product requirements and report results.

Act: undertake actions to continuously improve process performance.

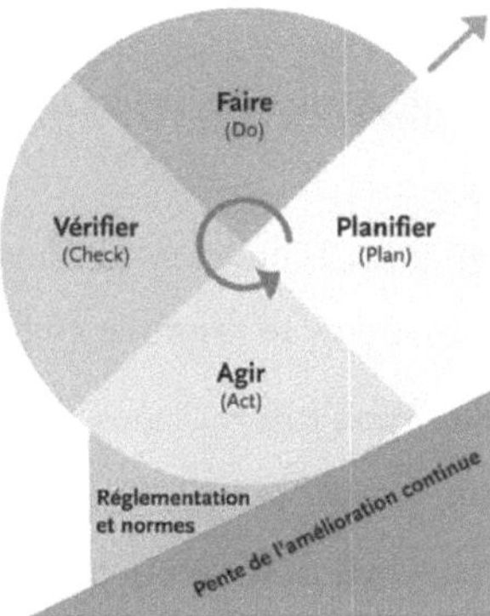

Figure 5: Process-based quality management system model

4.4 "ISO 22000" Standard

Food safety is concerned with the presence of hazards associated with food at the time of consumption (ingestion by the consumer). Because food safety hazards can be introduced at any stage of the food chain, it is essential that the entire food chain be adequately controlled. Therefore, food safety is ensured through the combined efforts of all actors in the food chain. Organizations involved in the food chain include feed and primary producers, food manufacturers, transport and storage operators and subcontractors, retail and food service outlets (as well as closely related organizations such as manufacturers of equipment, packaging materials, cleaning products, additives and ingredients). Service providers are also part of the food chain.

This International Standard specifies the requirements for a **food safety management system that** includes the following elements, generally recognized as essential, to ensure the safety of food at all levels of the food chain up to the final stage of consumption. This standard defines requirements to enable an organization to:

a. To plan, implement, operate, maintain and update a food safety management system designed to provide products that, in accordance with their intended use, are safe for the consumer;

b. Demonstrate compliance with applicable food safety legal and regulatory requirements;

c. Evaluate and assess customer requirements, demonstrate compliance with agreed upon food safety requirements to improve customer satisfaction;

d. Communicate effectively on food safety issues with suppliers, customers and stakeholders in the food chain;

e. Ensure compliance with its stated food safety policy;

f. Demonstrate this compliance to interested parties; and have its food safety management system certified/registered by an external body, or perform a self-assessment/self-declaration of conformity to this International Standard.

BIBLIOGRAPHY

a. Bruno Schiffers, Babacar Samb, and Jérémy Knops. "Principles of hygiene and management of sanitary and phytosanitary quality". COLEACP. Belgium; 2011.pp: 364

b. H. Greenfield and D.A.T. Southgate "Food composition data: production, management and utilization". InFood, Rome, 2007. pp: 319.

c. R.F. KAHRS. "General principles of disinfection". Rev. Sci. Tech. Off. int. Epiz. 1995, 14 (1), 123-142.

d. Olivier Bouteau. "From HACCP to ISO 22000: Food Safety Management". AFNOR Edition. 2ème edition. 2008.pp :352.

e. ISO/TS 22002-1. Technical specification. "Prerequisite programs for food safety: part 1: food manufacturing". 2009.pp: 28.

f. NF EN ISO 9001. Quality management system: Requirements. 2000.pp : 40.

I want morebooks!

Buy your books fast and straightforward online - at one of world's fastest growing online book stores! Environmentally sound due to Print-on-Demand technologies.

Buy your books online at
www.morebooks.shop

Kaufen Sie Ihre Bücher schnell und unkompliziert online – auf einer der am schnellsten wachsenden Buchhandelsplattformen weltweit! Dank Print-On-Demand umwelt- und ressourcenschonend produzi ert.

Bücher schneller online kaufen
www.morebooks.shop

info@omniscriptum.com
www.omniscriptum.com

Printed by Books on Demand GmbH, Norderstedt / Germany